Magical SCIENCE

Magic Tricks for Young Scientists

Eric Ladizinsky

Illustrated by
Dianne O'Quinn Burke

Lowell House House
Juvenile
Los Angeles

Contemporary Books
Chicago

To Lisa, who grows flowers in barren rock —E. L.

For Matthew and Djem, with love —D. O. B.

Designed by Tanya Maiboroda and Lisa Lenthall

Manufactured in the United States of America

ISBN: 1-56565-026-3
Library of Congress Catalog Card Number: 92-15789

10 9 8 7 6 5 4 3

what's magical about science?

There's a secret ancient mystics knew,
A secret this book reveals to you:
The greatest magic show in town
Is here and there and all around.
In every flower and flying dove
In twinkling stars way up above.
Nature's magic makes things go—
The laws of physics behind the show!

Since ancient times, men and women who have understood the workings of nature have often been considered to possess "magical powers." The ancient astronomers, for instance, knowing a little bit about the orbits of planets, could predict when and where planets would appear in the sky. Their predictions inspired awe in people who were less knowledgeable than they were. The ancient alchemists performed all manner of wondrous feats, simply because they studied and understood the properties of chemicals.

Imagine traveling back in time, perhaps a thousand years ago, taking with you things that are commonplace today—a television, a radio, a laser, a computer. What a powerful wizard you would seem! Imagine what an ancient Egyptian might say if he or she saw a car. "What makes it go?" they'd ask. "Where are the horses to pull it?"

Perhaps scientists and inventors truly are the greatest magicians of all, for they reveal to us the magic of the universe in which we live. In this book, you'll learn of many natural wonders and the scientific principles that underlie them. You'll learn how to use these simple scientific principles to create some amazing magic tricks.

The tricks are categorized by the particular aspect of the natural world they illuminate (for instance, gravity, mechanics, chemistry, and so forth). After learning how each trick works, you'll then learn the science behind the illusion. To help you learn the language scientists speak, look for the floating boxes of scientific expressions. They will help you understand why your magic tricks work.

It's important to understand that being a good magician takes more than just understanding how a trick is done. You must be an entertainer as well. You must help your audience to "suspend their disbelief"—to help them marvel at the magic of the world. So follow the Magician's Code below, if great magic is what you wish to show!

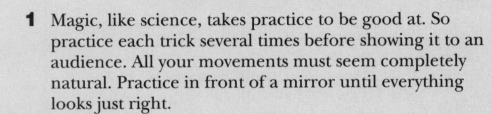

MAGICIAN'S CODE

1 Magic, like science, takes practice to be good at. So practice each trick several times before showing it to an audience. All your movements must seem completely natural. Practice in front of a mirror until everything looks just right.

2 Great magicians are also great storytellers. Capture your audience's imagination by telling them wonderful stories—by taking them on imaginary journeys to different times and places. For many of the tricks in this book, you'll find magical stories to tell. Use these, or if you prefer, invent your own.

3 To help your audience imagine you as a magician, it helps to dress like one! For a magical effect, you can use a top hat, a wand, a long-sleeve coat, or a flowing robe with a scarf wrapped around your head like a turban.

4 Make sure you have a show table to perform your tricks on.

5 NEVER reveal the secrets behind your magic. Magic, once understood, is no longer magical!

genie's fists

*Have you ever wanted to be Hercules
And move giant things as you please?
With just the strength of a feather
And by being really clever,
You can move the unmovable with ease.*

MAGIC

You Will Need

- volunteer
- assistant

Getting Ready

1 For this trick, you'll need a large assistant (perhaps an older brother or parent) to play the part of a giant stone Genie. He'll need to memorize a few lines before performance time. Have your assistant stand at your side during the trick.

Performing the Trick

1 Now it's time to take your audience to another time and place by telling them a story. Tell them, **"I once sailed with a gallant crew across the seven seas. During one violent storm, my mates and I became shipwrecked on a mysterious island in the Indian Ocean. When we awoke from the crash, we found ourselves in a magnificent golden palace whose door was guarded by a giant statue of stone."**

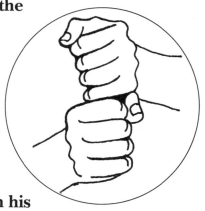

2 Have your assistant face the audience, extend his arms, put one fist on top of the other, and close his eyes. Then say, **"The statue was of a mighty sleeping Genie who stood with his**

fists stacked one on top of the other." Then say, "Suddenly the stone Genie opened its eyes and began to speak!"

3 Your assistant, playing the part of the Genie, should then say loudly, **"Unless you can break the bonds of my stone fists, you will be prisoners here forever!"**

4 Now say, **"The Genie had spoken. We would have to separate the giant stone fists to ever leave that place! Can any of you break the bonds of the fists?"** Ask a volunteer to come up. As your assistant presses his fists together with all his strength, ask the volunteer to try to pull the fists apart in the vertical direction. He or she will find it difficult or impossible to do.

5 Then say, **"The strongest of our crew also could not pull or pound the giant's fists apart. But in watching my mates' helpless struggle, I realized the answer."**

6 Have your assistant press his fists tightly together again. (Make sure his arms are fully extended.) Then slap two fingers of either hand against each of the two fists as shown, separating the fists, not crossing them.

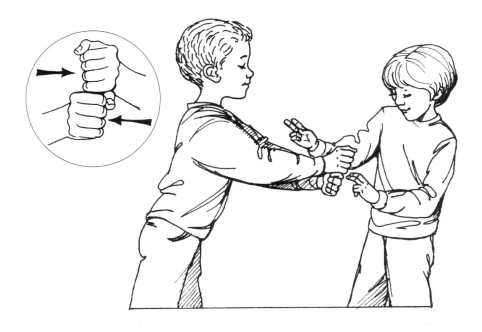

7 Then your assistant should say, **"You have learned a great secret—that the softest of blows can defeat the mightiest of foes. The golden doors are now open and you are free to leave. Farewell, magician!"**

SCIENCE

Why was it so easy to separate the fists? One way to understand why is to think about moving a friend on ice skates. It's a lot easier to push your friend forward than to lift her or him up into the air. Why? Because when trying to lift your friend up vertically, you have to fight a large force—the force of the earth's **gravity** pulling down on your friend.

When you push your friend forward horizontally, you don't have to fight gravity. The only force you have to fight is the friction of the blades against the ice, which is much less than the gravitational force and easy to overcome.

> **Gravity** is an invisible force field that all objects produce. Each object's force field attracts other objects around it. Small objects have very weak fields, but massive objects (like the earth) have very large fields.

When your assistant pressed his fists together, he pushed the top fist down and pulled the bottom fist up. All the force he generated was in the up/down, or vertical, direction. So, when your volunteer tried to pull his fists apart in that direction, he or she had to fight all this force. Since the assistant didn't push his fists together horizontally, very little force was generated in that direction. So, it was easy to knock his fists apart sideways.

electrical storm

From the lightning in storms
To the beeping of horns,
To a toaster's glow that warms

In every gadget that blinks
And every magnet that clinks,
It's charges behind these high jinx.

You Will Need

- wooden or plastic table (if the table you have available is metal, cover it with a tablecloth)
- tissue paper
- bowl
- drinking glass
- silk cloth

Getting Ready

1 Make sure the silk cloth is very dry.

2 Tear the tissue paper into very small pieces and put them in the bowl.

3 Wash and dry your hands thoroughly. (It's important that your hands are completely dry before performing this trick.)

4 Put the silk cloth, the bowl of tissue, and the glass (upside down) on your show table.

Performing the Trick

1 Begin this trick by telling your audience, **"On one of my travels into the remote mountains of the Himalayas, I discovered a remarkable tribe of people. They lived in one of the coldest places on earth, but all they had to wear was**

strange silk clothing that couldn't possibly protect them from the freezing storms."

2 Pick up the silk from the table and hold it in front of you. Then say, **"This is a piece of that special cloth. The tribe members told me it was made from magical silk spun by an enchanted worm—named Barney."**

3 Pick up the glass and rub it inside and outside with the silk cloth. While rubbing, say, **"They also told me that anything rubbed with the magical silk would be protected from the mightiest storms and the freezing cold."**

4 Rub the glass for two minutes or more. Be careful not to let your skin touch the glass. Then set the glass upside down on the table as shown (again without touching the glass).

5 Pick up the bowl of tissue pieces and take a pinch of them in one hand. Hold them 8" to 10" above the glass. Then say, **"And now, by rubbing this glass, I have made it magical, too! Any snow that touches the glass will be cast off."**

6 Sprinkle the tissue paper "snow" onto the glass and watch the magic. The tissue pieces will first stick to the glass and then fly off in an amazing way!

SCIENCE

Wow! What's going on? What strange invisible force is responsible for the tissue sticking to, and then flying off, the glass?

Well, not only flying tissue, but also nearly everything else you experience—from lightning to starlight to television—results from something called **charge**.

We know of two different types of charges: positive and negative. If two charged particles have the same charge (such as two negative charges), they will repel (push away from) each other. Oppositely charged particles will attract each other (come together).

> **Charge is a quality that some microscopic particles have. Charged particles create invisible force fields around them that can affect other charged particles. Moving charges create magnetic forces. Static (stationary) charges create electric forces.**

Most of the things around you—such as tables, trees, and air—contain about an equal number of negative and positive charges. Thus, the charges cancel each other out. That's why most of the time you don't notice their effect.

When you rubbed the glass with the silk, however, the glass "stole" (stripped off) negatively charged particles from the silk. The glass had more negative charges than positive, so it became negatively charged.

When you dropped the paper bits, the extra negative charges on the glass attracted the positive charges in the tissue bits (remember that opposite charges attract). This caused the paper to stick to the glass. Once the paper and glass touched, however, some of the extra negative charges on the glass leaked onto the paper. Can you guess why? They were trying to get away from each other (remember that like charges repel each other). So, the paper bits very quickly became negatively charged, too. Since the glass and the bits were both negatively charged, the paper bits flew off!

★grate vision

Have you ever seen a solid fence "disappear"?
Bet you have!

*If you've ever ridden in a car or on a bike
And happened to speed by a fence or the like,
Then just like Superman using X-ray vision,
The fence disappeared out of sight.*

MAGIC

You Will Need

- ruling pen (this is a special kind of pen used for drawing thin lines)
- tracing paper, 8½" x 11"
- book or newspaper with medium to large print
- ruler

Getting Ready

1 For this trick, you'll need to make a magic grating. Begin by laying the tracing paper down on a flat, smooth surface.

2 Now, using the pen and ruler, draw side-by-side (parallel) lines about ⅛" apart from each other all the way across the tracing paper.

3 Make a crisscross pattern by drawing more parallel lines ⅛" apart at a 90° angle to the lines already drawn.

4 Starting from the upper left corner and working toward the lower right corner, draw more parallel lines, this time on the diagonal of the paper. Don't forget to keep the lines ⅛" apart.

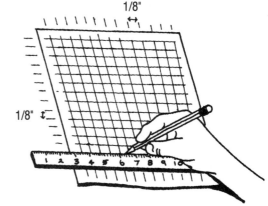

5 Finish your magic grating by drawing one last set of diagonal parallel lines from the upper right corner of the tracing paper down to the lower left corner.

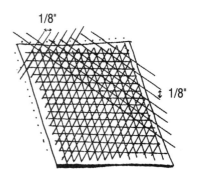

Performing the Trick

1 Ask the members of your audience to gather around your show table.

2 Place your book on the table and ask one volunteer to turn to any page.

3 With your magic grating, cover the page selected.

4 Ask your volunteers to try to read through the grating. They will find it impossible.

5 Now say, **"It is often said that a magician's hand is quicker than the eye. I have learned to move so fast that I can remove the grating, read what is underneath, and replace the grating—all in the blink of an eye."**

6 Slightly wet the fingertips of your right hand by licking them, then lean over the book and put your fingers over the lower right corner of the grating (if you're left-handed, put your fingers over the lower left corner). Tell your volunteers, **"You may see a little shaking of the paper between your blinking, but you won't see me remove it. Catch me if you can!"**

7 Now shake the grating back and forth over the page, using small, rapid movements (you only have to shake the grating about $1/4$" to $1/2$"). You will be able to see through the grating and read what's underneath. Amaze your audience by reading the covered page out loud.

SCIENCE

Why can you see through the grating when you shake it, but not when it's still? Why does the grating seem to disappear? Why does a fence seem to disappear as you speed by it on a bicycle? The anwer lies in many of the familiar objects around you.

Our eyes send messages to our brains that tell us what image we are seeing. Our brains hold on to those messages for a short time. So, even after an image is gone, your brain still sees it momentarily.

When something moves very fast, as with the grating in your trick, your brain does not have enough time to "see" it clearly. So your brain does something very interesting: it ignores what it can't see very well, especially if there is something easier to see lying in the same direction.

How is your grating like an electric fan?

A good way to understand what's going on in your trick is to use an analogy. That means using one idea you already understand in order to understand another idea. Consider an electric fan—it's very similar to your grating. When the fan is off, the wide blades block most of your view of the other side. Turn the fan on and the blades seem to melt away—you can see right through the spinning fan as if the blades weren't there. In fact, you can see entire objects sitting behind the fan, even though you know that only a tiny portion of the objects can be seen between the blades at any given time. Why is that?

As the blades rotate, the spaces between them also rotate. These rotating spaces allow you to see all the different parts of the objects behind the fan. Your brain then stores (remembers) all the parts and puts them together to form complete pictures. Meanwhile, it ignores what it can't see very well—the spinning blades—so all you see are the objects.

Your grating works the same way as the blades of a fan. When the grating is moved quickly, the tiny spaces between the lines reveal small bits of the page underneath. Your brain collects the bits of information, allowing you to construct a picture of the book underneath—which you can read!

fire trail

Parental Supervision Required

Have you ever heard of a combustion engine?

An engine makes every car go—
How it works is a thing you should know.
A neat trick with matches
Makes invisible gases
The secret behind the show!

You Will Need

- two wooden matches (wax coated)
- matchbox
- glass of water

Getting Ready

1 Put the matches, matchbox, and glass of water on your show table.

Performing the Trick

1 Look at your audience and say, **"One of the hazards of traveling back into the past is that I often end up in time periods before any modern conveniences were around—no electricity or hot water, no air conditioning or refrigerators. Try finding your way to the bathroom in the middle of the night in a dark medieval castle!"**

2 Now pick up the two matches and say, **"So, I developed a way to make a flame jump. Without leaving my bedside, I'd light one match and make the tiny flame jump from one candle to the next to light my way before getting out of bed."**

3 Light the two matches and hold one match over the other match as shown. Now say, **"Watch closely to see the flame jump."**

4 Blow out the bottom match. Make sure the rising smoke from the bottom match flows up into the match flame above it. The bottom match will magically light!

smoke ← trail

5 Blow out the bottom match two or three more times and let it re-light as before. (Just before you re-light it, say, **"Flame, jump!"**) Put both matches into the glass of water.

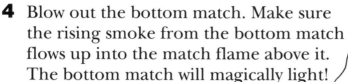

SCIENCE

Do you know why the match caught fire without being touched? Since it only happens when the smoke from one match flows into the flame of the other, the smoke must carry something with it. What does it carry?

When matches burn, chemicals in the wood and air react to produce heat and light. The matches also release chemicals, in the form of gas, into the air around them. In your trick, when you blew out the bottom match, this gas flowed up into the flame above. The gas became hot—so hot that it exploded, sending more heat down the smoke trail. That heat re-lit the bottom match.

Gasoline, sparks, and energy

When gasoline and air mix together, they also form a gas that explodes when heated (in this case by a spark). The energy from the explosion makes cars go.

So why are car engines called combustion engines?

It is interesting to note that your trick would not have worked with just any kind of matches. Not every burning object releases an explosive gas around it. It depends on what the object is made of. Gases that explode or burn when heated are **combustible**. That's why car engines are called combustion engines. And, you could say you performed a combustible trick!

change of mind

 As time goes on, people learn more and more about the brain and all the amazing things it can do. People just like you have invented complicated machines such as computers and space ships. Who knows what amazing things we will be able to do with our brains a hundred years from now? The possibilities are endless. With a little trickery and careful scientific observation, you can appear to have advanced mental powers right now. You can even appear to read minds!

You Will Need

- coin
- two chairs
- blindfold

Getting Ready

1 Place the blindfold and the coin on your show table, and put the chairs side by side next to the table.

Performing the Trick

1 Begin your trick by telling your audience, **"Not long ago, I took a trip through time—into the future. I found that the people of the future spend a lot of time developing their mental powers. They have even learned how to read the thoughts of others. I became good friends with the future people, and they taught me their secrets. Now I, too, have mastered their mind-reading techniques."** Then look at your audience, open your eyes really wide, raise your arms, and say, **"I will now give you a glimpse of the future."**

2 Ask two volunteers to come up and sit in the two chairs.

3 Give the coin to one of the volunteers.

4 Now tie the blindfold around your head, making sure to completely cover your eyes. Turn your back to the volunteers.

5 Tell your volunteers, **"The two of you must now secretly decide who shall hide the coin. Then, whoever has it must hold it in one hand."**

6 When the volunteers have decided, the magic begins. Ask both volunteers to close their hands into fists.

7 With your back still to your volunteers, tell them, **"Please hang your fists down by your sides, both of you. Now, whoever has the coin, raise the coin to your forehead. Remember, don't tell me who has the coin."**

8 Now tell your subjects, **"I want the person with the coin to concentrate really hard. Think of nothing else but the coin. The other volunteer must empty his mind. Let your thoughts wander."**

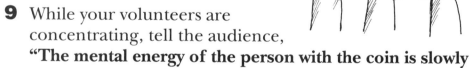

9 While your volunteers are concentrating, tell the audience, **"The mental energy of the person with the coin is slowly seeping into the coin. This 'brain energy' is strongest near the head. I will be able to see this invisible 'brain energy' because of my advanced mental powers."**

10 Tell both volunteers to put their fists on their knees, knuckles up.

11 Quickly remove your blindfold, turn around, and look at the backs of your volunteers' hands. Pretend to be thinking really hard. One person's hands will look the same. The other person's hands will look a little different from each other. One hand will seem darker than the other, and its veins will look larger and puffier. This was the hand hanging to the side. The lighter-colored hand, the one with smaller veins, is the one with the coin held to the volunteer's forehead.

12 Touch the lighter hand of the volunteer whose hands look different and say, **"Ahh, the coin is here and so are your thoughts!"**

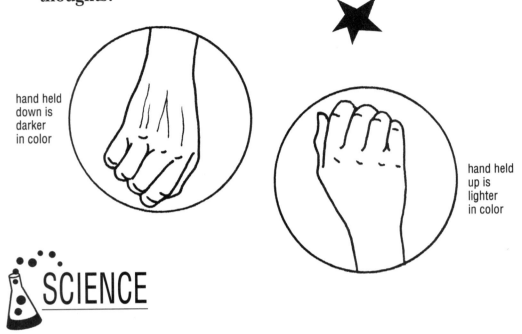

hand held down is darker in color

hand held up is lighter in color

SCIENCE

What's going on? Did "brain energy" really change the color of your volunteer's hands?

Well, it's really not brain energy. In fact, it's just our old friend gravity. Remember Genie's Fists (page 5)? The earth's gravity pulls everything down, including the blood in your body. The only thing that can get blood to flow up against gravity is your heart. Your heart acts as a little pump, pushing blood to all parts of the body, but it's harder for your heart to push blood up, fighting gravity, than down.

The hand that the volunteer held to his forehead was above his heart. That hand had less blood flowing to it than the hand hanging below his heart. The hanging hand was thus darker than the hand held to the forehead. The veins were also more swelled.

You can feel the effect of gravity with this simple experiment: turn upside down and feel the blood flowing "down" to your head!

on pins and needles

Parental Supervision Required

 Why do you need something sharp to cut apart something tough? Why can a pin puncture your skin? If you could see things microscopic, you'd have a handle on this topic. With sharp edges and sharp tips, it's smallness that does the trick!

MAGIC

You Will Need

- sharp needle (not too thin)
- 3"-long nail
- cork

- hammer
- goggles
- metal file

- two books of the same width
- quarter

Getting Ready

1 Put the needle into the cork so that the sharp tip sticks out about an inch from the bottom of the cork.

2 While wearing the goggles to protect your eyes, file the sharp end of the nail until it is almost, but not completely, flat.

3 Put the two books, the quarter, the needle and cork, the nail, and the hammer on your show table.

filed-down nail

Performing the Trick

1 Place the two books on the table side by side, leaving about a half-inch space between them. Place the quarter over the space as shown.

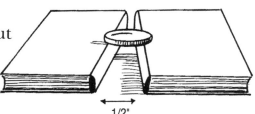

1/2"

19

2 Now look at the audience and say, **"It may be hard to believe, but at one time steel, glass, and plastic didn't exist. People eventually made all these things by mixing together all sorts of chemicals and minerals found in the air and ground. People are always trying to make things stronger and lighter by playing with different materials."**

3 Pick up the cork with the needle in it and show it to the audience. Say, **"When I was in the future, the future people gave me this needle. It is made of advanced materials and is stronger than our strongest steel. And I'll prove it to you."**

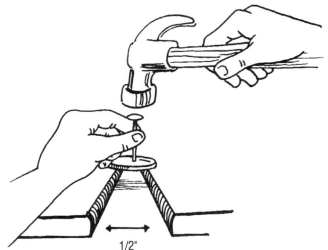

1/2"

4 Ask a volunteer to come up. Say, **"I want you to take this modern, but primitive, nail and try to gently tap it through this quarter by using the hammer."** Nothing will happen.

5 Center the coin between the books again and place the needle tip at the center of the coin.

6 Say, **"Now watch what happens when I use this little needle!"** Strike the top of the cork with the hammer (not too hard). The needle will go right through the coin!

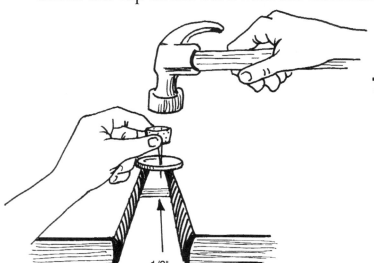

1/2"

7 Now exclaim, **"Wow! And this needle is many times smaller than the nail. Imagine what amazing things we could build with this kind of material!"**

 # SCIENCE

In reality, the nail and the needle are both made of steel. So why does one penetrate the coin, while the other doesn't?

You can understand this trick by referring to Friendly Float (page 31). If all the weight had been carried by one person, the floater would have been impossible to lift. But when all of your friends lifted the floater together, they distributed the weight. Each lifted only a small part of the weight, making the floater easy to lift.

The coin is made up of incredibly small particles called **atoms**. The atoms are stuck together by **chemical bonds**. These bonds can withstand only a certain amount of force before they break. When you placed the needle against the coin, it came in contact with relatively few bonds. That's because the needle is very small at its sharp tip. When the needle was hit with the hammer, each of the bonds on the coin that came in contact with the needle experienced enough force to break it, so the needle could go through the coin.

> A **chemical bond** forms when charged particles in different atoms interact, causing the atoms to stick together. Each place where an atom sticks to another is called a **bond**.

The tip of the nail was much larger, however, especially after the top was filed flat. It came in contact with a lot more bonds. This time, the force of the hammer blow was shared between a lot more bonds. Each bond thus received a much smaller portion of the force. The smaller force on each bond was not enough to break them, so the nail couldn't go through the coin. (Something similar happened in Friendly Float. In holding up a floater, each person can withstand only so much weight before they "break," like a bond, and the floater falls from their grip. When more people share the weight—that is, when there are more "bonds"—they can carry the weight between them. Their bonds withstand the force and don't break.)

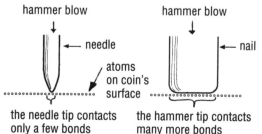

hammer blow   hammer blow

← needle   ← nail

atoms on coin's surface

the needle tip contacts only a few bonds   the hammer tip contacts many more bonds

balancing wand

Gravity

Do you think it would be hard to walk a tightrope in space? Tell you what—it would be easy, because there's nothing pulling you down in space. There's no gravity there. That's why the astronauts float. Here on earth, gravity is constantly pulling you and everything else down. And unless you have good balance, even walking isn't so easy (remember when you couldn't walk?). By making a magic wand just right, you can learn what it takes to walk upright.

You Will Need

- black construction paper
- white construction paper
- scissors
- masking tape
- ruler
- glue
- wooden dowel (approximately 10" long and ⅞" thick)
- two 1-oz. lead fishing weights

Getting Ready

1 Cut a strip of black paper 10" long and 2" wide.

2 Roll the paper strip tightly around the dowel as shown. Glue down the overlapping edge so the paper forms a tube. Until the glue dries, the tube can be held closed with a strip of masking tape.

3 When the glue is dry, remove the masking tape and slide the dowel out from the paper tube.

4 Now put the fishing weights into the tube. (If they are too large, pound them flat with a hammer.)

5 Cut two 2" x 2" pieces of white construction paper for the wand's tips. Neatly close and glue the ends of the black paper tube as shown.

Performing the Trick

1 Gaze into your audience and ask, **"Have any of you ever been to the planet Graviton? Well, you'd remember if you had. When I was there, all I did was fall down a lot. Graviton is circled by a moon called Nemesis, just like the earth is orbited by our moon. But Nemesis isn't small like our moon. It's huge. As it goes around Graviton, the direction of gravity keeps changing. It pulls down, then up, then sideways. Living on Graviton is a constant balancing act."**

2 Pick up the wand and hold it out. Then say, **"This wand is from Graviton. It has remarkable properties to deal with the constantly changing gravity."**

3 Now balance the wand on your finger as shown. (Make sure the weights are in the middle of your wand, or this won't work!)

4 Say to the audience, **"Now imagine Nemesis circling around above us to the east."** Hold the wand and slowly make a circle in the air to imitate Nemesis circling around. As you circle the wand, let the weights fall down to one end.

5 Ask a member of your audience to try to balance the wand the way you did. Your volunteer won't be able to do it.

6 Then say, **"The wand from Graviton has changed its balance point."** Lay the weighted end of the wand on the edge of a table as shown. The

weights will keep your balancing wand from falling over the edge.

7 Now pick up the wand and balance it on the tip of your finger!

Do you know why you could balance the wand on your finger? And why the wand did not fall over the edge of the table? Let's find out.

Try to balance a ruler on your finger. After a couple of tries, you'll discover that the ruler balances most easily when you hold your finger in the middle of it. That's the point where the ruler weighs the same amount on either side of your finger. This middle point is called the **center of gravity**. An object can be balanced at its center of gravity because the weight of the object has been equally distributed.

But what if one end of the ruler is heavier than the other end? In other words, what if its weight is not equally distributed? Because gravity works by pulling an object down at its heaviest point, the ruler would now balance at the heavier end. This time, the center of gravity is at that heavier end. If you had used a normal wand in your magic trick, you would not have been able to change its center of gravity. But by adding the fishing weights to the wand and moving them around, you could change it. The volunteer from your audience could not balance the wand on his finger as you did because you had changed the center of gravity—from the center of the wand to one end of it. With the wand's center of gravity now at one of its ends, you could balance it on the edge of the table!

glass blowing

Parental Supervision Required

The pushing of air to and fro
Causes vacuums that make the winds blow.
From a push on a bike racer's ride,
To great storms from which you would hide,
The filling of vacuums with air
Helps racers or makes storms that scare.

MAGIC

You Will Need

- candle
- candle holder
- two glasses of water
- book
- matches

Getting Ready

1 Set up one glass of water and the candle as shown. Make sure the wick is aligned with the center of the glass. If either the glass or the candle is too short, use a book to align their heights.

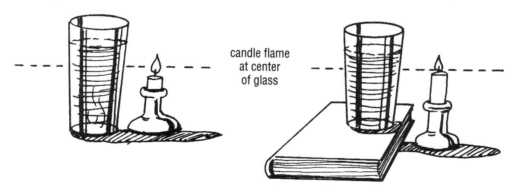

candle flame
at center
of glass

2 Now light the candle. Drop the blown-out match into the other glass of water.

3 Get on the side of the glass opposite the candle and blow at the middle of the first glass. You might blow the candle out. More likely, you will need to move closer or farther away from the glass while blowing at the middle of it to find the perfect distance needed to blow the candle out.

4 When you find the exact distance, remember it for your trick.

Performing the Trick

1 Set up the first glass of water and candle exactly as you did before.

2 Look at the audience and say, **"I once went to Venice, Italy, to learn the delicate art of glass blowing. I wandered through the streets looking for a teacher."**

3 Now light the candle, drop the match into the second glass of water, and say, **"I found a small shop that had 'GLASS BLOWING' written above the door. Inside, I met a strange old woman who told me that she did not teach how to blow beautiful shapes out of glass, but instead taught the ancient magic of blowing *through* solid glass."**

4 Get on the opposite side of the glass from the candle, just like you practiced before, and blow the candle out.

SCIENCE

How did the air you blew get through the glass and blow out the candle? Can you guess?

The trick is that the air blown on *your side* of the glass caused air to be sucked toward the *candle side* of the glass. How did that

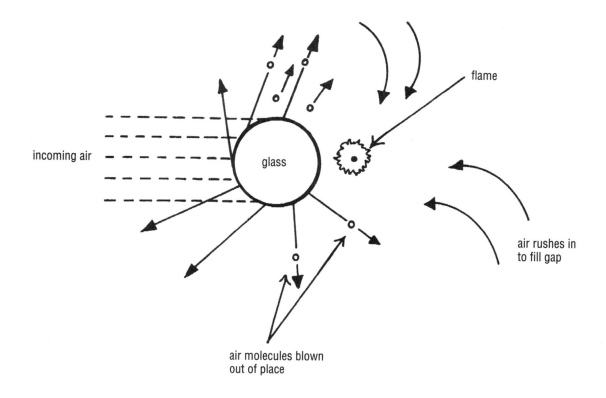

incoming air

glass

flame

air rushes in
to fill gap

air molecules blown
out of place

happen? Look at the picture above. When you blew on the glass, air molecules collided with the glass and were deflected outward as shown. As the deflected molecules moved away from the glass, they collided with and swept away other air molecules surrounding the glass. That left an area in back of the glass with fewer molecules. The air in back of the glass thus had a lower **density** of air.

Because the air in back of the glass had a lower density than the area in front of the glass, a small vacuum was created. Air around the glass then rushed into this area to replace the molecules that had been swept away. The rush of air blew out the candle.

Density is the amount of stuff in a certain volume of space. It's a measure of how tightly packed molecules are, whether they are molecules of air, water, glass, or stone. A **vacuum** is a region of space that has a very low density of gas.

A moving object, such as a car, will push air aside in the same way and cause air to rush in behind it. Race-car drivers, bicyclists, and long-distance runners will often get behind another racer to take advantage of the in-rushing air, which gives them a push and saves their energy. This strategy is called **drafting**.

eggstraordinary!

Why do some things float while other things sink?
Hmm, what do you think?
With some eggs and salt and H2O,
You'll find the answer in your magic show!

MAGIC

You Will Need

- large container of salt
- long-handled spoon
- large bowl
- pitcher of water
- pen
- three eggs
- three tall drinking glasses

Getting Ready

1 Fill the large bowl with three glasses of water.

2 Add salt to the water and mix. Keep adding salt until an egg will float in the bowl.

3 Fill one glass with pure water. Fill the second glass with saltwater from the bowl. Now fill the third glass half full of saltwater, then very slowly pour the pure water on top of the saltwater so that it flows gently along the side of the glass. If you pour slowly enough, the pure water will not mix with the saltwater.

4 Place the three glasses on your show table. Be careful not to shake them up! Then put the three eggs, pen, spoon, and the large bowl (emptied) on the table.

first pour saltwater into glass

then very slowly pour in the pure water

28

5 It is important that you know which glass is which. For your trick, you'll need to remember that an egg will sink in the glass of pure water, float on top of the saltwater, and float in the middle of the half-and-half glass.

Performing the Trick

1 Ask two volunteers to come up to your show table to assist you.

2 Have one of the volunteers use the pen to write different labels on the eggs. Have him or her write *Sink* on one egg, *Float* on another, and *Hover* on the third.

3 Now take the egg marked *Sink* and say, **"OK, now everybody concentrate. Think *sink*!"** Gently drop the egg into the glass filled with pure water. The egg will sink to the bottom.

4 Now take the egg marked *Float* and say, **"Concentrate again, please. Everybody think *float*!"** Drop the egg into the glass of saltwater. It will float!

5 Finally, take the egg marked *Hover* and tell your two volunteers, **"I want one of you to think *sink* and the other to think *float*. Let's see who has the more powerful magic."** Gently drop the egg into the half-salt/half-water glass. The egg will float in the middle of the glass. Look at them and say, **"You are equally powerful!"**

6 Pour the eggs and water into the bowl and challenge anyone else to repeat the trick.

SCIENCE

Do you know why the eggs floated in saltwater but sank in pure water? What's different about regular water and saltwater? By carefully observing what happens when you make the saltwater, you'll get a clue that will unravel the secret.

When you add salt to the water and mix it, it seems to disappear, but the level of the water in the bowl doesn't change (it doesn't rise or sink). So, you fit more stuff in the same amount of space. Whereas just the water filled the space before, now the salt *and* the water filled the space. Thus, the saltwater contains more stuff (molecules) in a certain **volume** than does regular water. A certain volume of saltwater, therefore, weighs more than an equal volume of pure water.

> The **volume** of an object is the amount of space it takes up, or occupies.

When you drop an egg in a glass of water (salty or pure), it can't dissolve the way salt can. It has to make room for itself. So, it pushes *a volume of water that's equal to its own size* out of the way. The sides of the glass don't allow the egg to push water sideways, so the only way it can push the water is up, against gravity. The water also has weight and pushes back.

If the egg weighs more than the volume of water it needs to displace, it sinks and the water level rises. If the water is heavier than the egg, the egg won't be able to push the water up. The egg then floats. By the results of your magic trick, you can see that the egg is heavier than pure water but lighter than saltwater.

Why did the egg labeled *Hover* float in the middle of the glass?

friendly float

One of the most amazing powers humans have is the ability to cooperate with one another. By working together, we can do things that seem impossible. The building of the great pyramids of Egypt and the rocket trips to the moon may seem like magic, but they are simply the result of many people sharing their ideas and working together.

By cooperating with some friends, you can make someone appear to float in the air by just touching him or her with your fingertips!

MAGIC

You Will Need

- four good friends

Getting Ready

1 Find a place with plenty of room, such as your basement, living room, or den. You can even perform this trick outside on the grass.

2 To become good at this trick, you'll need to practice it together with your friends a few times before performing it in front of an audience.

Performing the Trick

1 Gather your four friends together and choose the smallest one to be the "floater."

2 Tell the floater to stand very straight and stiff like a robot. If she is too relaxed, she will be hard to lift.

3 Now instruct your friends to move into the positions shown on the next page. One person should stand behind the floater and put his index fingers under the floater's arms. Two other friends should squat down and put their index fingers under the floater's heels. The magician stands in front of the floater with one finger under the floater's chin.

4 When everyone is ready, tell your audience, **"Friendship is one of the most magical things in the world. With our Friendship Power, we will raise the spirits and the body of our friend. She will float on an invisible sea of our good feelings. Now, everybody think good thoughts about our friend."**

5 Now tell your friends to lift together by counting, **"One, two, three, lift!"** It is important that the floater remains stiff and all four friends raise their fingers up together. Raise the floater as high as you can and lower her gently back to the ground.

SCIENCE

What's the secret behind the floating mystery? The best way to understand it is to try the following experiment. Take a small salad plate and try to balance it on one finger (be careful not to drop and break it!). It will balance when your finger is in the middle of the plate—that is, at its **center of gravity**. There is an equal amount of weight on one side of the center of gravity as on the other side. So, if you want to share the weight of the plate equally with someone else, both of you must hold it the same distance away from its center of gravity and on opposite sides of it. If four people want to share the weight equally, they must surround the plate's center of gravity. You can see this for yourself by asking three friends to help you hold up the plate as shown. Notice that by surrounding the center of gravity, you are sharing the plate's weight equally. (If one person were to move his finger closer to the 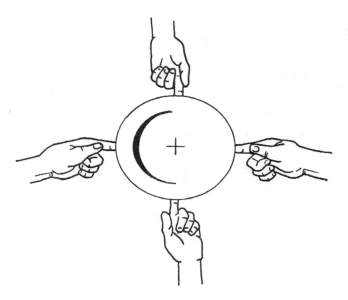 middle of the plate, that person would support less weight, and it would be harder for the other three to hold up the plate.) This is what happened in your magic trick.

Your floater's center of gravity is approximately in the middle of her body. The lifting points were chosen carefully; they surrounded her center of gravity so all four people could equally share her weight. With each of you having only one-quarter of the floater's weight to lift, you could use only your fingertips to make her float in the air!

the rising tide

Parental Supervision Required

Have you ever wondered where great storms come from? Where do they get the power to lift millions of gallons of water miles into the sky—water that eventually falls to the earth as gentle rain? In a very small way, you can create the same invisible force that powers storms, re-creating the magic of nature.

You Will Need

- cork
- three wooden matches
- matchbox
- dinner plate
- tall, thin drinking glass (wide enough to fit over the cork)
- pitcher of water

Getting Ready

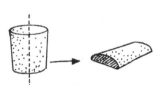

1 Cut the cork in half, lengthwise, from the top of the cork to the bottom. Ask your mom or dad to help you do this.

2 Stick the three wooden matches into the cork as shown, making your own little Viking ship.

3 Place the Viking ship, the dinner plate, the tall, thin glass, and the pitcher of water on your show table.

Performing the Trick

1 Now it's time to tell the audience a story. Say, **"It is often said that long before Columbus discovered America, ancient Viking ships landed there. But the Vikings did not have the giant wooden sailing ships that Columbus had. They had only tiny ships to sail the dangerous waters. How could they have**

survived the long, treacherous storms of the wild seas?" Pause dramatically, then say, **"Now it's time to tell the tale."**

2 Now pick up the pitcher of water and say, **"Here I have the water of the oceans."** Pour the water onto the dinner plate until the water is about ¼" deep in the middle of the plate.

3 Show your audience the cork with the matches in it. Tell them, **"This cork is made from the wood of an ancient Viking ship. This magical wood can do amazing things."**

4 Place the cork on the water and say, **"Imagine this is a tiny Viking ship at sea. When the sailors would see a storm approaching, they would light the torches and go below deck."** Light the matches on the cork and drop the match you used to light them in the pitcher of water.

5 Pick up the glass and say, **"When the storm got really bad, the torches were blown out. Then the magic wooden ship cast a spell on the water."**

6 Slowly lower the glass over the little Viking ship until it hits the bottom of the plate. Gently hold the glass down until the matches go out and the water under the cork rises. Once the water rises as high as it will go, gently let go of the glass. The water will stay high in the glass!

7 Then say, **"This is the secret of the Viking sailors: When the torches went out, the magic wood drew up a giant column of water. This allowed the boat to sail safely above the greatest waves and survive the storm!"**

SCIENCE

What's happening? How could the water rise when gravity is constantly pulling it down? Is this really magical wood?

You can begin to understand what's going on by remembering the Change of Mind trick (page 16). What pushed your blood up to your head, fighting gravity? Your heart did, acting as a little pump. Something like a pump pushed the water up, too—only this time, the pump is invisible. Can you guess what it is? Here's a clue: it's all over the place. It's the air!

Air secret

The air is made up of a huge number of molecules that are constantly banging into one another like miniature billiard balls. The **temperature** of the air tells us how fast the molecules are moving. The hotter the air, the faster the molecules move. The cooler the air, the slower the molecules move. A source of heat, like a flame or the sun, can give energy to slow-moving molecules and speed them up. It can turn cold air into hot air.

As you lowered the glass over the Viking ship, the air inside the glass got warmer from the burning matches. The molecules of air inside the glass then started moving very fast, and a lot of them could escape from the glass before it was completely lowered. This caused the air in the glass to become less dense.

When the glass completely covered the cork, the matches went out because they used up all the oxygen (they need a constant supply of oxygen to keep burning). Meanwhile, the water outside the glass was being pushed down by all the air molecules banging into it. The air inside the glass was pushing down, too, but there were fewer molecules in the glass than before. Remember, some escaped when it was hot. That means the density of the air inside the glass was less than the density of air outside of it. So, the air outside the glass pushed more than the air inside did, forcing the water in and up the glass.

The sun heats the air in the same way to produce storms. In heating the air, the sun causes the air's density to lower, and wind is the result. Why? Because cool, high-density air rushes into an area of warm, low-density air. If powerful enough, wind (especially the wind in storms) can lift water up and out of the ocean.

dance of the butterflies

What makes the flame bright in candlelight? What makes a car go? Do you know? All around—in the sea and the ground, and in the air here and there—are chemicals that, when mixed, create amazing natural tricks!

In this magic trick, you'll produce a chemical reaction that will cause audience attraction!

You Will Need

- empty jam or jelly jar
- large cork
- sheet of thin cork
- funnel
- powder that effervesces (like Alka-Seltzer)

- glue
- sharp pencil
- scissors
- bowl
- spoon
- pitcher of water

- sheet of notebook paper
- thin, brightly colored tissue paper

Getting Ready

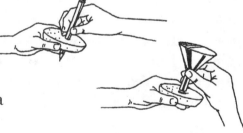

1 Take the pencil and push completely through the middle of the cork to make a hole.

2 Push the small end of the funnel through the hole as shown. Make sure it's a tight fit or your trick won't work.

3 Lay a sheet of tissue paper over the butterfly at right, and carefully trace its outline. Use the scissors to carefully cut out your butterfly.

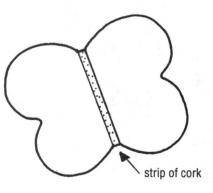

strip of cork

4 Repeat step 3 with different-colored tissue paper to make three or four butterflies.

5 For each butterfly, cut a thin strip of sheet cork ⅛" wide and 1" long. These will be your butterflies' bodies.

6 Glue the strips to the middle of the tissue butterflies as shown on the previous page. Let dry.

7 Place 10 seltzer tablets in the bowl and crush them with the spoon until you have a fine powder.

8 Put the jar, the cork with the funnel, the butterflies, the pitcher of water, and the bowl with the powder on your show table.

Performing the Trick

1 Look at the audience, smile, and say, **"I recently visited a magical, faraway land where everything was alive—rocks and socks, fishes and dishes, popcorn, candy, and anything handy. Everything was alive and dancing."**

2 Carefully pick up the tissue butterflies, show them to the audience, and say, **"I made friends with these beautiful butterflies in the magical land. They were so sad when I told them I'd be leaving, so I took them with me. But when I took them away from their magical home and friends, they stopped dancing and turned into mere paper butterflies."**

3 Put the butterflies back on the table and pick up the pitcher of water. Pour the water into the jar until it's half full. While pouring the water, say, **"This is some water that I brought back from the magical place."**

4 Now pick up the bowl of powder and say, **"And this is some of its magical white soil."** Pour the powder into the jar. The water will begin to bubble.

5 Quickly plug the top of the jar with the cork and funnel.

6 Pick up the butterflies, bring them close to your mouth, and say, **"I've brought you a taste of home, little ones, so you can dance once more."**

7 Place the butterflies into the top of the funnel. They will rise up and appear to fly around. Now tell your audience, **"Their magic has returned. Look, they're dancing!"**

 SCIENCE

Do you know why the butterflies suddenly "came to life"? Let's find out.

All the substances around you are made up of atoms. There are many different kinds of atoms, and they combine in many different ways to form all the neat stuff around you. The water in your jar is made of two types of atoms: hydrogen and oxygen. The white powder is made of different kinds of atoms, too. When you poured the powder into the water, it started a **chemical reaction**. This means the water and the powder exchanged some of their atoms. That resulted in a group of atoms called carbon dioxide splitting off from the powder in the form of gas. The carbon dioxide gas then traveled through the water to the top of the jar.

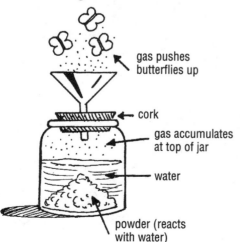

gas pushes butterflies up

cork

gas accumulates at top of jar

water

powder (reacts with water)

As more and more carbon dioxide gas accumulated in the top of the jar, pressure built up and a stream of carbon dioxide gas shot through the funnel. That pushed the paper butterflies up. As long as the water and powder reacted to produce the gas, the butterflies appeared to dance.

mini movie

Have you ever wondered how a movie works? To find out, let's make a thaumatrope. The real magic in this trick takes place in one of the most mysterious places on earth—your brain!

You Will Need

- 2" x 2" piece of stiff, white cardboard
- 8½" x 11" sheet of white paper (unruled)
- magic marker or pen

- glue or rubber cement
- scissors
- two 8" pieces of string (or thread)
- large needle

Getting Ready

1 Put the paper on top of the pictures below and trace the circles, the bird, and the cage. Be sure to mark the X's.

2 Cut out the two paper circles with scissors.

3 Trace around one of these circles on the white cardboard and cut it out.

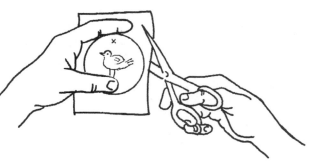

4 Glue the paper circle with the bird onto the cardboard circle and let it dry.

5 With the needle, poke a hole through one of the X's on the cardboard circle. This will make it easy for you to correctly line up the cardboard circle with the other paper circle.

6 Glue the second paper circle to the other side of the cardboard, making sure that the cage isn't upside-down and that the X is over the needle hole.

7 Use the needle to thread the pieces of string or thread through both sets of X's.

8 Tie the two ends of each piece of string together to keep them from slipping back through the holes. You've now completed your thaumatrope.

Performing the Trick

1 Show the thaumatrope to your audience. Invite a volunteer to come up and ask if he or she knows a way to get the bird inside the cage. Your volunteer will say that it's impossible, because the bird is on one side of the thaumatrope, and the cage is on the other.

2 Now tell your audience, **"This little bird used to be in the cage, but she escaped when I wasn't looking. If we could travel back in time, we could see the bird where she used to be—in the cage."**

3 Hold the strings in each hand, as shown. Twirl the thaumatrope around and around until the strings are tightly wound.

4 Then say, **"Back through time we flow. Watch the little bird go!"** While everyone is watching, pull on the strings quickly. As the thaumatrope unwinds, the bird will appear in the cage!

SCIENCE

Do you know why the bird appeared to be in the cage when the thaumatrope unwound? Do you remember the grating you made in Grate Vision (page 11) and the way your brain stores (remembers) a picture of what you see for a short time?

This is what happened in your magic trick when the thaumatrope unwound. Your audience continued to see the bird even after the image was gone. Here's the important point: By the time the cage appeared, your audience was still "seeing" the bird and was fooled into thinking the bird was in the cage. Their brains had *superimposed* the new cage image with the lingering bird image. Then, by the time they saw the real bird again, they superimposed it with the lingering cage image, and so on. Scientists call this **persistence of vision**.

The same thing happens at the movies (or "moving pictures"). If you look at a strip of movie film, you will see that it is actually a series of still pictures, one after the other. When each picture is flashed onto the movie screen, our brains remember it and merge it with the next picture. In this way, the series of still pictures appears to be moving.

the bird and cage images superimposed

So, in a way, you did time travel. You saw things the way they were just a moment ago—in the past!

dinosaurs didn't dance

All around you, at every moment, there is information in the air waiting to be decoded. Music and pictures from faraway lands, the voices of people from all over the world, and the secrets of the night sky all exist in the form of **electromagnetic waves**. These are the kind of invisible waves detected by radios, televisions, and telescopes. Is it ever possible to escape these mysterious waves? By performing a magical feat, you can make the waves retreat!

MAGIC

You Will Need

- two shoe boxes with lids
- aluminum foil
- black construction paper
- scissors
- glue
- green fern leaf (or other fragrant plant)
- battery-powered transistor radio
- paints or markers
- paintbrush

Getting Ready

1 Use the brush to coat the inside of the first shoe box with glue. Then place a large sheet of aluminum foil in the shoe box, carefully pressing it into the corners and along the edges of the box. Let it dry, then cut off the excess foil around the edges.

2 In the same way, line the lid of the first shoe box with foil.

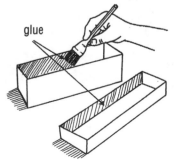
glue

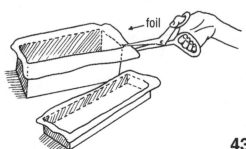
foil

3 Now glue the black paper to the aluminum foil in the first box and lid. Be sure to completely cover the foil and cut off any excess black paper. Let dry.

black paper

4 Rub the fern leaf against the black paper until the box smells like fern inside.

5 Now take the second shoe box and lid and glue just the black paper to the inside of both.

6 In big letters, paint or write *Prehistoric* on the box with aluminum foil in it, and write *Modern* on the box without aluminum foil. Add any designs you'd like to the boxes.

7 Put the shoe boxes and the radio on your show table.

Performing the Trick

1 Start by choosing a volunteer from your audience. Then point to the two boxes on the table and say, **"I have two shoe boxes. One is rather ordinary; the other is quite extraordinary. I took one of them back into time—to the Age of the Dinosaurs."**

2 Pick up the box labeled *Prehistoric* and say, **"While standing in a misty prehistoric jungle filled with giant insects and terrifying dinosaurs, I opened this magic box and filled it with the time and space around me. Though I returned to the present, the inside of this box is still in the past, 100 million years ago."** Open the box, show the inside to your audience, then ask your volunteer, **"Can you smell the ferns of that ancient jungle?"**

3 Now put down the shoe box and turn on the radio so everyone can hear music playing. Say, **"The music you hear is made by humans using electronic devices such as this radio."**

4 Then show the inside of the *Modern* box to your audience, put the radio inside, and close the lid. You will still hear the music. Say, **"The music still reaches your ears because this box contains things from our time, including radio signals."**

5 Now take the radio out of the *Modern* box and lower it into the *Prehistoric* one. Before closing the lid, say, **"But the inside of this box is from the Age of the Dinosaurs, long before humans made radio waves of music. There are no signals in this box for the radio to receive."** Close the lid and the music will stop!

Do you know why the music stopped in one box and not the other? Remember the difference between the shoe boxes? The one with the aluminum foil in it stopped the music, and the other didn't. The secret is in the aluminum.

You can understand what's going on by using an analogy.

Bobbing corks and wiggling particles

Even though radio waves are invisible, you can imagine what they're like by thinking about a cork floating in a pool. When a wave passes by the cork, what does the cork do? It bobs up and down. In this analogy, the waves in the pool represent radio waves, and the cork represents a charged particle inside a radio antenna. When radio waves hit charged particles on a radio's antenna, they cause the particles to bob up and down—to wiggle. (This is how radio waves are decoded into music. Information carried on the radio waves causes the charged particles in the antenna to wiggle in very precise patterns. Then, the electronics in the radio read these wiggle patterns and cause a speaker to vibrate with the same patterns as the wiggles. In this way, the speaker produces the sounds you hear.)

There are charged particles in the aluminum you used in your trick, too. These particles are free to move about, and they moved in response to the incoming radio waves in such a way as to neutralize them. Suddenly, your radio's antenna had no more information to decode. The charged particles in the antenna could no longer wiggle.

The *Modern* box didn't shield the radio, because charged particles in cardboard, unlike in the aluminum, are not free to move about and neutralize radio waves.

falling reign

There's something invisible here and there—
On earth, it's nearly everywhere.
It constantly presses on all around.
In fact it's the reason we hear every sound.
So let's use the push of the unseen air
To amaze our friends while they sit and stare!

MAGIC

You Will Need

- thin magic marker pen (blue is best)
- pitcher of water
- small drinking glass
- scissors

- piece of clear plastic (the plastic must be larger than the mouth of the glass and thin enough to be cut with scissors, but not too thin or it will deform under the weight of water)
- blue food coloring

Getting Ready

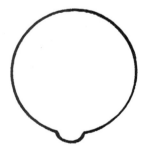

1 For this trick, you'll need to cut out a plastic circle that's a little bit larger than the mouth of the glass and that has a tab on it. Do this by laying the glass upside down over the plastic, drawing around the glass to make a circle, then cutting just *outside* of the pen line. Be sure to leave a little tab as shown.

2 To give the water some color, add a few drops of food coloring to the pitcher and stir.

3 Put the glass and the pitcher on your show table.

4 Place the plastic disk in your right hand as shown (or your left hand if you're left-handed). Practice holding it so it can't be seen. (This is called palming.)

Performing the Trick

1 Using your left hand, pour the water into the glass until it begins to overflow. (Remember, the plastic disk is being palmed in your right hand.)

2 Look at the audience and say, **"Oh look—I've never known when to stop pouring. I've also never had enough sense to come in out of the rain. It's so inconvenient getting wet, so I invented a way to keep the wet rain away."**

3 Put down the pitcher, then reach for the glass with your left hand, looking like you're about to pick it up. As you do this, reach over to the glass with your right hand. Keeping the plastic disk hidden, cover the mouth with your hand and secretly

palm disk

place the disk over the top. Make sure it covers the top evenly. It will stick to the mouth of the glass if centered properly.

4 Pick up the glass with your left hand, placing your other hand over the disk. Turn the glass upside down quickly, but gently. Keep the mouth of the glass tilted slightly away from the audience so they can't see the plastic.

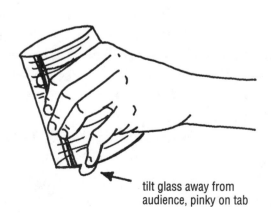

tilt glass away from audience, pinky on tab

5 Now, slowly remove your right hand. The disk will remain stuck to the glass and the water won't spill out.

6 Hold the glass over your head and say, **"I don't need an umbrella, do I!"** Then secretly push on the little tab sticking out with your pinky. The plastic will come off and the water will fall on your head!

 SCIENCE

The plastic is holding the water in, but what's holding the plastic up? What's fighting gravity?

The answer is all around you. It's the air. The earth's gravity holds air molecules around it and doesn't allow them to escape into space. This air forms the earth's atmosphere. The weight of all that air pressing down on you is called **air pressure**. On a microscopic scale, air pressure on an object is the force caused by air molecules constantly colliding against the object. These collisions happen in every direction, even up.

The force of air molecules pounding up against the bottom of the plastic disk is greater than the weight of the water above the plastic disk. So the plastic disk doesn't fall. Though the water's weight may seem to be the stronger force, the air actually is!

★glassical music

What's in your ear that allows you to hear?
Did you know it's a kind of drum?
The little drum vibrates
When air near you gyrates
To produce the music you hum.
From a singing cricket to a big drum's pound,
The shaking of molecules makes every sound.
So out in space or on the moon,
There'd be no music or song of loons
Cause there isn't any air
Way out there.

MAGIC

You Will Need

- two identical crystal wineglasses
- water
- thin steel wire (a little longer than the diameter of the glass)
- eyedropper
- pliers

Getting Ready

1 Fill both wineglasses about one-quarter full with water and place them about 10" to 12" apart on your table.

2 With the pliers, bend the steel wire at the ends and lay it over the mouth of one glass.

3 Now hold the base of the second glass and dip a finger in the water to wet it slightly. Slowly and lightly rub

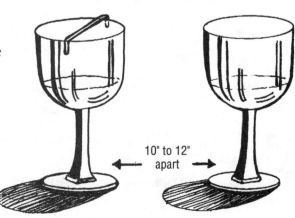

10" to 12" apart

the wet finger around the rim of the second glass. As your fingers follow the rim, the glass will begin to vibrate and make a high-pitched noise. When your finger becomes dry, wet it again and continue rubbing the glass.

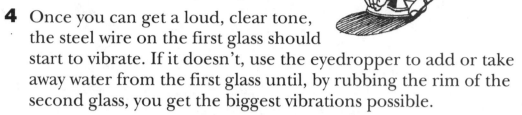

4 Once you can get a loud, clear tone, the steel wire on the first glass should start to vibrate. If it doesn't, use the eyedropper to add or take away water from the first glass until, by rubbing the rim of the second glass, you get the biggest vibrations possible.

5 Set up the wire and the two glasses of water on your show table, in the exact positions needed.

Performing the Trick

1 Look at the audience and say, **"Long ago, one hundred knights in shining armor made a solemn vow to a powerful wizard. They promised to protect their village from a great beast that the wizard said would one day come and destroy them."**

2 Pick up the steel wire and hold it in front of you. Then say, **"The wizard gave each knight a sliver of magic steel and told them that when the day came, the slivers would unite to make a great sword. This sword would kill the beast."**

3 Now say, **"But the hundred knights were confused. They asked the wizard how they'd know when that day had arrived."** Lay the wire over one glass and say, **"The wizard replied that they'd**

know when their slivers began to dance upon a goblet of wine. He would summon them by making his own goblet sing. Then, no matter where in the world the knights were, their magical slivers would dance to the music of the wizard's goblet."

4 Put your finger in the water and rub the glass without the wire. The wire will jump. Then open your eyes wide and say, **"*I* was that wizard. And the time has come for the slivers to unite!"**

 SCIENCE

Can you guess what's happening? Have you ever felt the floor shake when a stereo hit just the right note? Have you ever seen windows rattle as a big jet flew overhead? Would you believe it's merely the air that's shaking things around?

When you rubbed the first glass, you caused the atoms in the glass to vibrate back and forth. The vibrating atoms hit the air molecules surrounding the glass, which caused the air to vibrate. The vibrating air molecules next to the glass then collided with air molecules farther away. Just like dominoes, one hitting the next, the vibration traveled away from the first glass to the second one. The vibration also traveled to your ears. (That's the tone you heard!) When the vibrating air collided with the second glass, it caused the glass to vibrate, shaking the metal wire along with it. But why did the second glass shake? Because it had the same **natural frequency** as the first glass.

> **Frequency** refers to how fast something vibrates back and forth. **Natural frequency** is the frequency at which an object will vibrate if provided energy (for instance, if it is struck).

Most objects, like the glass, have a natural frequency. If the frequency of the vibrations from the first glass match the natural frequency of the second glass, the second glass will vibrate. A similar thing happens when a swing is pushed at the right moment—it rises higher and higher. If the first glass vibrates at a different natural frequency than the second glass, not much will happen. By making the water level in the second glass just right, you made its natural frequency match the frequency of the vibrations created by the first glass.

sands of time

Have you ever wondered how we humans tell what stuff is made of? Look at an ingredient label sometime. Even simple substances can contain a lot of different things (or ingredients). Some ancient Indian magic can help you to understand some ideas scientists use to figure out what things are made of.

You Will Need

- bowl
- dinner plate
- large glass mixing bowl
- blue food coloring
- three small glasses
- pitcher of water
- paraffin wax

- cheese grater (with a fine grating on it)
- thin plastic tube, 8" to 10" long and ¾" in diameter (available at hardware stores)
- red, white, and blue sand (available at craft stores)

Getting Ready

1 Add blue food coloring to the pitcher until the water is dark blue.

2 Fill each of the three glasses with a different color of sand.

3 Using the finest grating possible, grate the paraffin wax over the bowl until you have a good supply of paraffin shreds.

4 Take a small handful of the red sand and mix it with some paraffin shreds. Roll the mixture in your hands like clay to make a little cake that can fit loosely inside the plastic tube. Use just enough paraffin to make the sand cake solid.

5 Repeat step 4 for both the white and the blue sand. It is important to make the cakes different shapes so you can tell which is which by just feeling them. Put the cakes into the plastic tube.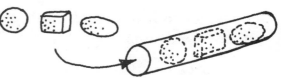

6 Put the tube, the mixing bowl, the three glasses of sand, the pitcher of blue water, and the dinner plate on your table.

Performing the Trick

1 Look at the audience and say, **"Once while traveling through ancient India, I came upon a strange and beautiful crystal-blue lake."** Pick up the pitcher and pour the blue water into the bowl until it's three-quarters full.

2 Now say, **"And standing in this strange lake was an even stranger old man with long white hair and crystal-blue eyes. He told me this lake was fed by three mighty rivers. Each river brought sand from faraway places, which mixed here to create this magic lake."**

3 Slowly pour the white sand into the bowl. While pouring the sand, say, **"The first river brought sand from a place of great white mountains—the place where the old man was born."**

4 Now pour in the red sand and say, **"The second great river brought sand from a wild red desert land—the place where the old man learned his magic."**

5 Finally, slowly pour in the blue sand and say, **"And the third mighty river brought sand from a silent blue forest, where the old man found understanding and peace."**

6 Now, carefully put the plastic tube into the bowl. *Once the tip is underwater,* let the sand cakes slip out into the bowl.

7 Mix the sand together with the tube. *Stir gently so as not to break the sand cakes and ruin the trick!*

8 Now stop stirring and say, **"Then the old man did an amazing thing. He**

reached into the lake and pulled out each color of sand separately. Like so!"

9 Reach your hand into the mixing bowl and lift out each of the sand cake shapes you made, one by one. Then crush each of them with your hand and sprinkle them onto the plate. As you sprinkle the white sand cake, say, **"The old man said, 'This is the land in which I was born—the time of my childhood.'"** With the red sand cake, say, **"This is the red land of my youth—the time of learning and adventure.'"** With the blue sand cake, say, **"Finally, the old man said, 'This is the land of memory—the time of peace and understanding—the place I go to now.'"** (Note: As you lift each sand cake out of the water, keep your fingers spaced like a rake so any mixed sand passes out of your hand into the water. Before you take your hand completely out of the water, close it around the sand cake.)

10 Look at the audience and say, **"The old man then smiled, bid me farewell, and sank into the quiet blue of the lake."** Let audience members come up and try to separate the sand!

SCIENCE

Scientists figure out what various things are made of by finding ways to separate their ingredients. For example, by heating saltwater to 100° Celsius, you can find out what it's made of. The water boils away but the salt remains. So, heating separates the saltwater into its two ingredients: salt and water. The idea of separating something into its parts was important to your trick.

In molding some of the sand into sand cakes, you gave the cakes a property that made them different from the regular sand. Namely, they were larger than the grains of sand, so they could not slip through your fingers. You also gave the sand cakes a property that made them different *from each other*: their shape. By giving each colored sand cake a different shape, you could tell which color cake you lifted out of the mixing bowl.

By using these differences between the sand cakes and the regular sand, you could separate the white, red, and blue sand in the mixing bowl—the old man's magical ingredients!

★ x-ray vision ★

Do you ever wonder why scientists use so much complicated junk? Did you know that it's because most of what happens in the universe—from giant explosions in space to the dancing of tiny atoms—is completely invisible to the naked eye? Your five senses show you only a small part of the world. Only by creating special instruments can we probe the invisible world. In this trick, you'll use an instrument to "see" through solid matter.

MAGIC

You Will Need

- four small (fairly strong) magnets
- small toy compass
- glue
- Scotch tape
- black construction paper
- scissors
- four small, empty boxes (such as individual-size raisin boxes)
- white paint
- colored paint (any color will do)
- small paintbrush
- aluminum foil (the heavy kind works best)

Getting Ready

1 Open both ends of the raisin boxes, paint the outsides white, and let dry.

2 Using the other color, paint the numbers 1, 2, 3, and 4 on your boxes. Let dry.

3 Glue a magnet to the inside of each as shown. The box with #1 on it should have a magnet glued above the #1, inside the box. The #2 box should have a magnet glued to the right of the #2. The magnet glued in box

#3 should be glued below the #3. Finally, the magnet for box #4 should be glued to the left of the #4. Note: Before the glue dries, put the compass on each box and make sure the arrow *tip* points to each magnet. If the *tail* of the arrow points toward the magnet, spin the magnet around in its place until the compass tip points to the magnet. Then let the glue dry.

4 Glue the ends of the boxes shut and let dry.

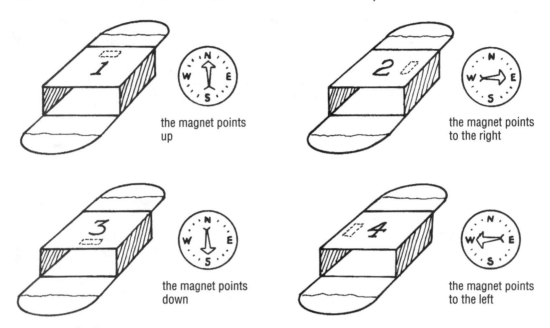

the magnet points up

the magnet points to the right

the magnet points down

the magnet points to the left

5 Now line the blocks up side by side, with the numbers right side up. Scan the compass across the middle of the boxes and watch what happens. The needle points up on box #1, to the right on box #2, down on box #3, and to the left on box #4. You can tell which block you're over by the way the compass points!

6 With the black construction paper, make a tube that's about 6" long and wide enough so that the compass fits snugly inside.

7 Take a large sheet of aluminum foil and lay it over the boxes, bending it over them to make a cover. Paint an arrow on the foil.

8 Put the boxes and the aluminum cover on your show table. Also put the tube on your table, but make sure the audience can't see the compass that's inside.

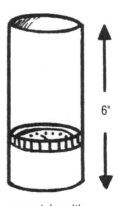

6"

paper tube with compass inside

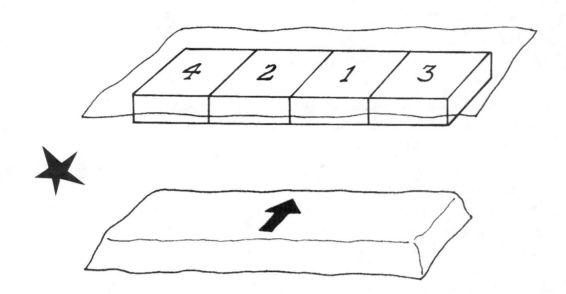

Performing the Trick

1 Have a couple of volunteers come up to assist you.

2 Turn your back to the volunteers and tell them to put the blocks right side up, side by side, in any order they like. They should not tell you what the order is. Then have them cover the boxes with the aluminum foil, with the arrow on it pointing right side up.

3 Now say, **"People are always trying to make things simpler and smaller. Calculators that used to be as big as rooms can now fit in your pocket. Consider X-ray machines, too. Right now they're big, complicated machines, but in the future, they'll look like this—an X-ray machine from the future!"** Show the tube to your audience without letting anyone see the inside of it.

4 Lean over the table and put the tube over the middle of the aluminum cover on the far left side and look into it (don't bring your eye all the way to the tube, or you'll block all the light entering it). You will see the compass needle move. If it points up, the block underneath is #1. If it points right, it's #2, and so on. As you move over the different blocks with the tube, the needle will change position. Shout out the order of the blocks underneath to your audience.

Any guesses why you knew the positions of the letters without "seeing" them? The magnets inside the boxes produced an invisible field that went right through the foil and moved the compass needle in your "X-ray machine." But where does the strange invisible force of the magnets come from?

Well, inside the magnets are charged particles. These negatively charged particles are called **electrons**. An electron creates an invisible force field around itself, called an **electric field**. When electrons move, they create another kind of field, this one called a **magnetic field**. Most materials contain many electrons. Since the electrons move around in many different directions, the magnetic fields they each create point in different directions, too. As a result, they cancel each other out (this is like two people pushing in opposite directions on a wagon—it won't go anywhere).

In magnetic materials, all the charged particles inside are spinning in exactly the same way so that all the little magnetic fields add up to produce a single, large force field (like many people pushing on a wagon in the same direction). This is what is happening with each of the magnets in the boxes.

The compass needle is also magnetic. The spinning electrons in the needle are affected by the invisible magnetic fields of the magnets. As a result, the compass needle moves around as it passes over each magnet until it lines up with that magnet's spinning electrons.

The world's biggest magnet

Can you guess why all compass needles point north? Would you believe it's because the earth is a giant magnet? Amazingly enough, deep beneath the earth's surface, trapped in an immense flowing river of molten rock, are charged particles circling around the earth's axis. The moving charges create an invisible magnetic field that spans the entire planet. It is this gigantic field that you "see" with a compass needle. Think of it—a toy compass is like a window to the inside of the earth!

hole milk

Parental Supervision Required

Most audience mind traps are set
When they believe what they see, they get.
So here's a trick where they'll extol
There's some milk when there's really a hole.

MAGIC

You Will Need

- tape measure
- clear drinking glass
- scissors
- milk
- sheet of newspaper
- masking tape
- pin or needle

- plastic epoxy (use glue that dries clear)
- two 8½" x 11" sheets of clear plastic acetate (¹⁄₃₂" thick)
- hammer
- nail

Getting Ready

1 To perform this trick, first make a plastic tube that fits loosely inside the glass. Do this by taking two measurements with the tape measure: (1) the circumference of the glass, and (2) its height (H). Subtract ¼" from the height measurement.

circumference

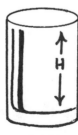

height

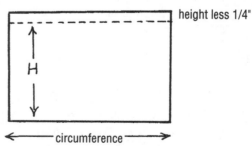

height less 1/4"

circumference measurement

2 Now cut a piece of plastic, using the two measurements for the height and width. (Example: if your glass measures 5" tall and 9" around, cut out a piece of plastic 4¾" x 9".)

3 Now roll the plastic into a tube, overlapping the edge by ¼". Glue the plastic ends together, using the masking tape to hold them shut while the glue dries.

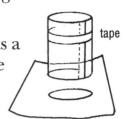

4 Now cut out a circular piece of plastic to use as a cover for the plastic tube. Do this by laying the tube end over another plastic sheet, drawing around it with a marker, then cutting it out.

tape

5 Using the hammer and nail, punch a small hole (about ¼" across) in the middle of the cover. Then neatly glue the cover to one end of the plastic tube and let it dry.

6 Now test the plastic tube for leaks. Fill it with water and cover the hole with your finger. If there are leaks, use a little glue to fill them in.

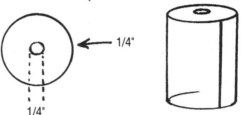

7 Fill the glass about ⅕ full with milk. Then, with your finger over the hole, lower the plastic tube into the glass (open end down). The milk will be forced up into the space between the glass and the plastic, and the glass will appear full of milk (if it doesn't, add more milk). Slide your finger off the hole, but keep pushing down on the plastic. The milk will return to its original level.

8 Make a cone out of your sheet of newspaper and tape it shut.

milk's level with hole uncovered

milk's level with hole covered

Performing the Trick

1 Tell your audience to excuse you for a moment while you get a glass of milk. Put the plastic tube into the milk glass to make the glass look full. Return to your audience.

2 Say to your audience, **"Have you ever wondered where all the stuff that magicians make disappear goes? Well, there are invisible holes in space that lead to another place. If I can find one of these invisible holes, and I've made this cone just right,**

then I'll funnel the milk through the hole to that other world out of sight."

as you pretend to pour, slide finger off hole

3 Pick up the newspaper cone with your free hand. Tilt the milk glass into the cone as if you were going to pour the milk in, but don't pour! Instead, secretly slide your finger sideways off the hole. The milk will settle to its real level, and it will look as if you've poured most of the milk into the cone.

4 Put the glass down, hold the cone with both hands, and walk toward your audience as if nervous about spilling the milk.

5 As you walk, say **"I think I see a hole!"** Run toward someone and pretend to trip, letting the cone fall on his or her head!

 # SCIENCE

There are two secrets to this trick. One involves air pressure and the other has to do with human psychology.

When your finger was over the little hole and you lowered the plastic into the glass, the milk pushed on the air in the tube. But that air was trapped—by the plastic on all sides and by the milk on the bottom. So, the air pushed back on the milk. The air pressure inside the tube was high enough, in fact, to push the milk up into the space between the glass and the plastic. When you took your finger off the hole, the milk encountered no resistance when it pushed on the air inside the tube. That's because that air now had a place to go—out of the hole. So the milk went into the tube.

Even though your glass was never full, the audience assumed it was when they saw milk around the sides. They compared it to something familiar—a full glass of milk—and they assumed this glass was the same without really knowing for certain. Most magic tricks work because an assumption is made on the part of the audience. The audience is fooled into believing something is true when it isn't. By contrast, good scientists try to avoid making conclusions about something before they have directly observed it.

which way is north?

Parental Supervision Required

Electromagnetism

No matter where in the world you are, in a South American jungle or an Arabian desert, one thing is always the same—the direction a compass needle points. Think of it—anywhere on earth, all compass needles point to the North Pole. What draws them northward? The answer involves one of nature's grandest mysteries. The same invisible electromagnetic fields that steer a compass light the stars!

MAGIC

You Will Need

- 1' of thin electrical wire (16 gauge)
- 12' of thick electrical wire (12 gauge)
- dark, long-sleeve shirt or jacket
- scissors
- D-size battery
- compass
- electrical tape
- large Band-aid
- wire cutters (long-nose pliers can also cut wire)
- red food coloring (optional)

Getting Ready

1 Using the scissors, strip about ½" of the plastic insulation off the ends of the thin wire to expose the bare wire.

2 Next wrap the thin wire around your right index finger, forming a coil. (Use your left index finger if you are left-handed.) Don't make the coil too tight—you want to be able to take it on and off easily.

3 Twist the wire at the base of your finger. That will help the coil stay together. Tape the wires down at your palm.

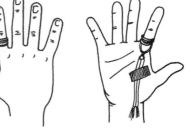

4 To hide the coil from your audience, wrap the large Band-aid around it. To prevent any suspicion, you can squirt a tiny amount of red food coloring underneath the Band-aid so it looks like you have a cut on your finger.

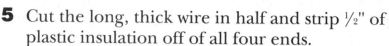

5 Cut the long, thick wire in half and strip ½" of plastic insulation off of all four ends.

6 Twist each bare end of the thin wire (the coiled one on your right hand) to a bare end of the thick wire, as shown. Then wrap electrical tape around the twisted connections.

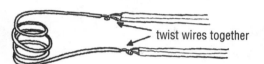

twist wires together

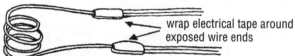

wrap electrical tape around exposed wire ends

7 Put on your long-sleeved shirt and feed the two long wires up your right sleeve across your chest and out of your left sleeve into your left hand.

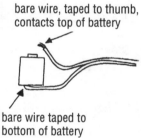

bare wire, taped to thumb, contacts top of battery

bare wire taped to bottom of battery

8 Tape one bare, thick-wire end to the bottom of the D-size battery. Tape the other end so that the exposed wire lays across your left thumb. This way, when holding the battery in your left hand, you can easily bring your thumb to the battery and touch the wire end to the battery top at will.

9 Put the compass on your show table. Remember, the coil is on your right index finger, the battery is in your left hand, and you connect the battery to the coil by pressing the wire on your left thumb against the top of the battery. Make sure the audience sees only the back of your right hand.

battery

wires under shirt

Performing the Trick

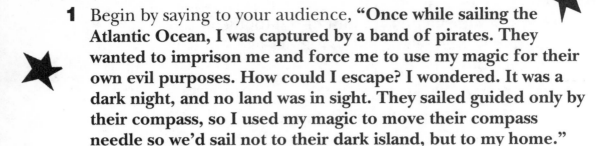

1 Begin by saying to your audience, **"Once while sailing the Atlantic Ocean, I was captured by a band of pirates. They wanted to imprison me and force me to use my magic for their own evil purposes. How could I escape? I wondered. It was a dark night, and no land was in sight. They sailed guided only by their compass, so I used my magic to move their compass needle so we'd sail not to their dark island, but to my home."**

2 Put your right hand near the compass and move it around. Then say, **"Take me home!"** and bring your left thumb to the battery top. The compass needle will follow your right hand.

3 Then say, **"Return to north!"** and take your left thumb off the battery top. The compass needle will again point to the north. Touch the battery again to show your control over the compass.

 SCIENCE

What's going on? Remember X-Ray Vision (page 55) and the way the magnets made the compass needle move?

In your trick, the battery, by a chemical reaction, created an excess of negative charges that wanted to fly away from each other. But the charges couldn't move until they had a path to flow through. When the wire on your thumb was connected to the battery, the excess negative charges got the needed path. They flowed through the wire, just like water flows through a pipe.

Then what happened? When the negative charges reached the coil, they flowed around and around, creating a magnetic field (just as the electrons circling the atoms in the magnets did in X-Ray Vision). It was this magnetic field that caused the compass needle to move.

The flow of charges through a wire is called electrical current.

When you disconnected the battery, the excess negative charges it produced again had no path to flow through. So, the charges stopped flowing and the magnetic field disappeared.